THE RADIOLOGICAL INCIDENT
IN HUEYPOXTLA

The following States are Members of the International Atomic Energy Agency:

AFGHANISTAN	GERMANY	PALAU
ALBANIA	GHANA	PANAMA
ALGERIA	GREECE	PAPUA NEW GUINEA
ANGOLA	GRENADA	PARAGUAY
ANTIGUA AND BARBUDA	GUATEMALA	PERU
ARGENTINA	GUYANA	PHILIPPINES
ARMENIA	HAITI	POLAND
AUSTRALIA	HOLY SEE	PORTUGAL
AUSTRIA	HONDURAS	QATAR
AZERBAIJAN	HUNGARY	REPUBLIC OF MOLDOVA
BAHAMAS	ICELAND	ROMANIA
BAHRAIN	INDIA	RUSSIAN FEDERATION
BANGLADESH	INDONESIA	RWANDA
BARBADOS	IRAN, ISLAMIC REPUBLIC OF	SAINT KITTS AND NEVIS
BELARUS	IRAQ	SAINT LUCIA
BELGIUM	IRELAND	SAINT VINCENT AND
BELIZE	ISRAEL	THE GRENADINES
BENIN	ITALY	SAMOA
BOLIVIA, PLURINATIONAL	JAMAICA	SAN MARINO
STATE OF	JAPAN	SAUDI ARABIA
BOSNIA AND HERZEGOVINA	JORDAN	SENEGAL
BOTSWANA	KAZAKHSTAN	SERBIA
BRAZIL	KENYA	SEYCHELLES
BRUNEI DARUSSALAM	KOREA, REPUBLIC OF	SIERRA LEONE
BULGARIA	KUWAIT	SINGAPORE
BURKINA FASO	KYRGYZSTAN	SLOVAKIA
BURUNDI	LAO PEOPLE'S DEMOCRATIC	SLOVENIA
CAMBODIA	REPUBLIC	SOUTH AFRICA
CAMEROON	LATVIA	SPAIN
CANADA	LEBANON	SRI LANKA
CENTRAL AFRICAN	LESOTHO	SUDAN
REPUBLIC	LIBERIA	SWEDEN
CHAD	LIBYA	SWITZERLAND
CHILE	LIECHTENSTEIN	SYRIAN ARAB REPUBLIC
CHINA	LITHUANIA	TAJIKISTAN
COLOMBIA	LUXEMBOURG	THAILAND
COMOROS	MADAGASCAR	TOGO
CONGO	MALAWI	TONGA
COSTA RICA	MALAYSIA	TRINIDAD AND TOBAGO
CÔTE D'IVOIRE	MALI	TUNISIA
CROATIA	MALTA	TÜRKİYE
CUBA	MARSHALL ISLANDS	TURKMENISTAN
CYPRUS	MAURITANIA	UGANDA
CZECH REPUBLIC	MAURITIUS	UKRAINE
DEMOCRATIC REPUBLIC	MEXICO	UNITED ARAB EMIRATES
OF THE CONGO	MONACO	UNITED KINGDOM OF
DENMARK	MONGOLIA	GREAT BRITAIN AND
DJIBOUTI	MONTENEGRO	NORTHERN IRELAND
DOMINICA	MOROCCO	UNITED REPUBLIC
DOMINICAN REPUBLIC	MOZAMBIQUE	OF TANZANIA
ECUADOR	MYANMAR	UNITED STATES OF AMERICA
EGYPT	NAMIBIA	URUGUAY
EL SALVADOR	NEPAL	UZBEKISTAN
ERITREA	NETHERLANDS	VANUATU
ESTONIA	NEW ZEALAND	VENEZUELA, BOLIVARIAN
ESWATINI	NICARAGUA	REPUBLIC OF
ETHIOPIA	NIGER	VIET NAM
FIJI	NIGERIA	YEMEN
FINLAND	NORTH MACEDONIA	ZAMBIA
FRANCE	NORWAY	ZIMBABWE
GABON	OMAN	
GEORGIA	PAKISTAN	

The Agency's Statute was approved on 23 October 1956 by the Conference on the Statute of the IAEA held at United Nations Headquarters, New York; it entered into force on 29 July 1957. The Headquarters of the Agency are situated in Vienna. Its principal objective is "to accelerate and enlarge the contribution of atomic energy to peace, health and prosperity throughout the world".

THE RADIOLOGICAL INCIDENT
IN HUEYPOXTLA

INTERNATIONAL ATOMIC ENERGY AGENCY
VIENNA, 2022

COPYRIGHT NOTICE

For further information on this publication, please contact:

Incident and Emergency Centre
International Atomic Energy Agency
Vienna International Centre
PO Box 100
1400 Vienna, Austria
Email: Official.Mail@iaea.org

© IAEA, 2022
Printed by the IAEA in Austria
September 2022

IAEA Library Cataloguing in Publication Data

Names: International Atomic Energy Agency.
Title: The radiological incident in Hueypoxtla / International Atomic Energy Agency.
Description: Vienna : International Atomic Energy Agency, 2022. | Includes bibliographical references.
Identifiers: IAEAL 22-01529 | ISBN 978–92–0–136222–3 (paperback : alk. paper) | ISBN 978–92–0–136322–0 (pdf)
Subjects: LCSH: Radiation — Safety measures. | Radioactive waste disposal | Radioactive substances — Theft — Hueypoxtla (Mexico). | Nuclear industry — Security measures.
Classification: UDC 614.876 (72) | IAEA/RAD/INC

FOREWORD

The use of radioactive material is beneficial to medicine, research and industry all over the world. However, the transport of radioactive material may give rise to the risk of incidents with the potential for radiological exposure that could impact the safety of people, property and the environment. Even with precautions in place, nuclear safety and security related events that do not directly target the transported radioactive material may occur.

As part of its activities dealing with the safety and security of radioactive material, the IAEA follows up on events involving radioactive material to provide an account of the circumstances of the event and relevant medical, environmental, safety and security aspects. Organizations with responsibilities for radiation protection and the safety and security of radioactive material can also learn from the account.

A radiological incident occurred in Mexico in December 2013. A vehicle transporting a teletherapy unit head with a Category 1 cobalt-60 radioactive source from a medical facility in Tijuana to a disposal facility in Santa Maria Maquixco, Mexico, was stolen by a group of armed individuals. A number of individuals who came into contact with the radioactive source received doses in excess of the effective dose limit specified in Mexican regulations.

This publication describes actions taken by Mexican authorities in response to this incident and dose assessments of the individuals exposed to the radioactive source.

The IAEA is grateful for the assistance of the National Commission for Nuclear Safety and Safeguards, Mexico, in preparing this publication and thereby sharing its experience with other Member States. The IAEA officers responsible for this publication were M. Hussain and K. Smith of the Incident and Emergency Centre.

EDITORIAL NOTE

CONTENTS

1. INTRODUCTION

1.1. BACKGROUND

This publication describes the radiological incident that occurred in Hueypoxtla, Mexico, in December 2013. The incident involved the theft of a radiotherapy machine head containing a ^{60}Co radioactive source[1] (Category 1 radioactive source with an activity of 95.24 TBq at the time of the incident) resulting in radiation exposure to members of the public. This publication provides information about the circumstances of the incident and the response actions taken by the Mexican authorities at the national level.

This publication also describes the actions taken by the IAEA's Incident and Emergency Centre (IEC), upon notification of the incident, in performing its response role in case of a nuclear or radiological emergency and in order to meet the obligations as per the Convention on Early Notification of a Nuclear Accident (Early Notification Convention) [1] and the Convention on Assistance in Case of a Nuclear or Radiological Emergency (Assistance Convention) [2].

Paragraph 1.14 of IAEA Safety Standards Series No. GSR Part 7, Preparedness and Response for a Nuclear or Radiological Emergency [3], states that "The requirements apply for preparedness and response for a nuclear or radiological emergency in relation to all those facilities and activities, as well as sources, with the potential for causing radiation exposure, environmental contamination or concern on the part of the public warranting protective actions and other response actions". Paragraph 1.16 of GSR Part 7 [3] states that "The requirements apply for preparedness and response for a nuclear or radiological emergency irrespective of the initiator of the emergency, whether the emergency follows a natural event, a human error, a mechanical or other failure, or a nuclear security event".

IAEA Safety Standards Series No. SSR-6 [4] establishes the specific arrangements to be made for the safe transport of radioactive material. IAEA Safety Standards Series No. SSG-65 [5] provides guidance on the emergency preparedness and response arrangements to be made for the transport of radioactive material. IAEA Nuclear Security Series No. 9-G (Rev. 1) [6] provides guidance on the security considerations for the transport of radioactive material.

A simple, logical system for ranking radioactive sources into five categories has been provided by the IAEA in IAEA Safety Standards Series No. GSR Part 3 [7], in the Code of Conduct on the Safety and Security of Radioactive Sources [8], in IAEA Safety Standards Series No. RS-G-1.9 [9] and EPR-D-Values 2006 [10]. A ^{60}Co (Category 1) radioactive source was involved in this incident.

Table 3 of RS-G-1.9 [9] provides descriptions of the source categories in plain language. With regard to the risk in being close to an individual source for Category 1 sources, Ref. [9] states that: "This source, if not safely managed or securely protected, would be likely to cause permanent injury to a person who handled it or who was otherwise in contact with it for more than a few minutes. It would probably be fatal to be close to this amount of unshielded radioactive material for a period in the range of a few minutes to an hour." With regard to the risk in the event that the radioactive material in the source is dispersed by fire or explosion for Category 1 sources, Ref. [9] states that: "This amount of radioactive material, if dispersed, could possibly, although it would be unlikely, permanently injure or be life threatening to persons in the immediate vicinity. There would be little or no risk of immediate health effects to persons

---

[1] The term 'source' in this publication means 'radioactive source'.

beyond a few hundred meters away, but contaminated areas would need to be cleaned up in accordance with international standards. For large sources the area to be cleaned up could be a square kilometer or more".

IAEA Safety Standards Series Nos GSG-2 [11], GS-G-2.1 [12], GSG-11 [13] and GSG-14 [14] provide specific guidance on emergency preparedness and response for a nuclear or radiological emergency.

This incident has already been described in GSG-11 [13] as a case study in the context of arrangements for the termination of a nuclear or radiological emergency. For the completeness and consistency of information in response to this incident, some of the relevant details contained in GSG-11 [13] have been reproduced in this publication.

1.2. OBJECTIVE

Since 1988, the IAEA has produced publications on reported nuclear and radiological emergencies in the form of its accident reports and has provided support and assistance under the Assistance Convention [2] upon request by the States in which the accidents occurred. The primary objective of such publications is to identify and share lessons, and to improve the existing arrangements to prevent the occurrence of similar events. Twenty one (21) accident reports has been published by IAEA, chronologically, Ventanilla, Peru [15], Chilca, Peru [16], Fukushima, Japan [17], Lia, Georgia [18], Nueva Aldea, Chile [19], Cochabamba, Bolivia [20], Bialystok, Poland [21], Gilan, Iran [22], Samut Prakarn, Thailand [23], Sarov, Russia [24], Lilo, Georgia [25], Istanbul, Türkiye [26], Yanango, Peru [27], Tomsk, Russia [28], Tammiku, Estonia [29], San José, Costa Rica [30], Nesvizh, Belarus [31], Hanoi, Vietnam [32], Soreq, Israel [33], Dan Salvador, El Salvador [34] and Goiania, Brazil [35]. The findings and conclusions in these publications have provided an objective basis for learning lessons aimed to improve safety and emergency preparedness and response arrangements.
Similar to the incident discussed in this publication, previous events involving teletherapy sources have occurred in Thailand [23], Türkiye [26] and Brazil [35].

The objective of this publication is to compile and disseminate information about (a) the circumstances that led to the incident in Mexico, (b) the initial response by the national authorities and the notification to the IAEA, (c) the response activities of the IAEA and (d) the source recovery activities and follow-up actions. This publication seeks to help States to identify similar or precursor situations and take the necessary actions to either prevent incidents and accidents from occurring or to mitigate the effects of radiation injuries in a timely manner.

The information in this publication is intended for use by Member States' competent authorities, regulatory bodies, emergency response planners, first response organizations, law enforcement agencies and a broad range of specialists, including medical specialists, physicists and persons responsible for radiation protection and security of radioactive material including in transport operations which involve shippers, carriers and receivers, as well as facilities that use radioactive material.

1.3. SCOPE

This publication gives an account of the events leading up to, and following the incident, as well as the response actions taken. The publication includes the findings, conclusions and lessons identified from this incident.

Detailed information regarding diagnosis and treatment for most exposed individuals is not in the scope of this publication. At the time of drafting of this publication, medical treatment of the most exposed individuals was ongoing. Such information may be made available in the future as a separate publication or as an addendum to this publication.

Terms are used in this publication as defined in GSR Part 7 [3] and the IAEA Safety Glossary [36].

1.4. STRUCTURE

Information about the legal and regulatory framework in Mexico is provided in Section 2. In Section 3, the circumstances which led to the incident are described. Section 4 presents details on the response actions taken at the national level regarding the search and recovery of the source, public communication, medical management, radiation dose assessment and criminal investigations. Section 5 outlines the IAEA's actions in response to the incident. Observations, lessons identified, and good practices are presented in Section 6. Conclusions are presented in Section 7. The press releases issued by the IAEA are provided in the Appendix.

2. LEGAL AND REGULATORY FRAMEWORK IN MEXICO

In Mexico, the legal and regulatory framework with regard to radiation protection and the safety and security of radioactive material is based on the Political Constitution of the United Mexican States from which a series of laws, regulations and standards are derived. Article 89 of the Constitution, Fraction I, empowers the President of the Republic to "promulgate and execute the laws issued by the Congress of the Union, providing the administrative support for allowing its exact observance" [37]. The Federal Executive Branch, through the Secretariat of Energy, regulates and supervises compliance with the provisions on nuclear safety and radiation protection matters, attribution based on Article 33, Fraction X of the Organic Law of Federal Public Administration [37].

Pursuant to Article 17 of the Organic Law of the Federal Public Administration, the Mexican State Secretariat is authorized to delegate authority to subordinated administrative bodies to provide more effective attention and more efficient dispatch of matters of their competency [37]. These organizations need to have specific powers to resolve matters within the territory and competence determined in each case, in accordance with the applicable legal provisions. From an administrative standpoint, this article supports the creation of the national commission for nuclear safety and safeguards in Mexico, "Comisión Nacional de Seguridad Nuclear y Salvaguardias (CNSNS)".

Article 27 of the Constitution Law on Nuclear Matters (Nuclear Law) [37] establishes that nuclear energy has to be used only for peaceful applications and that use of nuclear fuel for the generation of nuclear energy, as well as regulation of its application in all areas, falls within the purview of the Mexican State.

2.1. ORGANIZATION OF CNSNS

CNSNS was established in 1979 with the responsibility to regulate the safety of nuclear installations in Mexico. CNSNS, located in Mexico City, reports to the Department of Energy. The mission of CNSNS is to ensure safety of the general population from facilities and activities[2] within the country, through the establishment of regulations and surveillance of the fulfilment of legal requirements and international treaties regarding nuclear safety, radiation protection and nuclear security. CNSNS's position within the government structure is presented in Fig. 1.

The Nuclear Law stipulates that CNSNS has to forward its opinion to the Department of Energy on siting, design, construction, operation, modification, suspension of operations, definitive shutdown and decommissioning of nuclear installations, prior to authorization by the Department of Energy (Article 50, Fraction IV) [37]. In this context, the legislation assigns responsibility to CNSNS for the following:

- Development of standards and regulations;
- Licensing and authorization of facilities and activities;
- Radiological environmental surveillance;
- Audits, technical visits and inspections to nuclear installations and radiation facilities;
- Examination and licensing of operating organizations of nuclear reactors;

---

[2] Facilities and activities is a general term encompassing nuclear facilities, uses of all sources of ionizing radiation, all radioactive waste management activities, transport of radioactive material and any other practice or circumstances in which people may be subject to exposure to radiation from naturally occurring or artificial sources [3].

- Inspections and audits relating to physical protection, nuclear security and safeguards;
- Licensing of imports, utilization, transport and storage of radioactive material;
- Licensing of radioactive waste repositories;
- Participation in international cooperation agreements;
- Research and development projects;
- National registry of occupationally exposed workers.

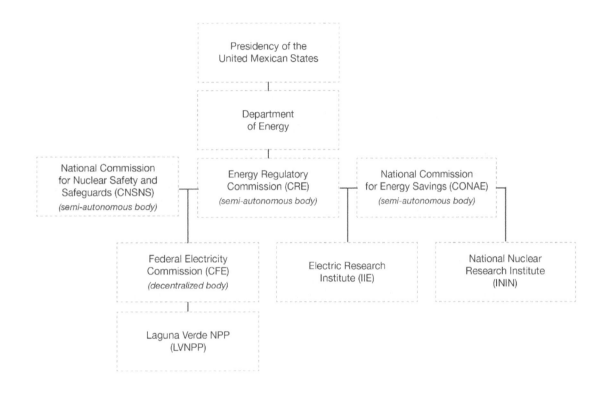

FIG. 1. CNSNS position within the government structure

2.2. LICENSING AND AUTHORIZATION PROCESS

CNSNS establishes or adopts regulations and guides upon which its regulatory actions are performed. For example, CNSNS adopted the regulations established by the United States Nuclear Regulatory Commission (U.S.NRC) for the Laguna Verde nuclear power plant. For radiation safety and security, CNSNS establishes general requirements that are based on the IAEA Safety Standards Series and Nuclear Security Series publications. All regulations and guides are issued by order of the Secretary of Energy based on CNSNS proposals.

The persons or organizations which intend to use ionizing radiation for non-medical diagnosis, industrial and research purposes or the provision of services, are required to submit an application for authorization to CNSNS.

2.3. REGULATORY REVIEW AND ASSESSMENT

Articles 17, 29, 32–35 and 50 of the Nuclear Law assign responsibility to CNSNS for regulatory review and assessment of submissions for authorization. The review and assessment of an applicant's[3] and licensee's[4] submissions are performed on the basis of a graded approach, and the potential magnitude and nature of the hazards associated with the facility, activity or practice are taken into account.

CNSNS verifies that the information contained in the submissions is accurate and sufficient to enable confirmation of compliance with the regulatory requirements in achieving the required level of safety and security of the radioactive material.

2.4. INSPECTION AND ENFORCEMENT

CNSNS inspection and enforcement activities are covered by Articles 36, 37, 38, 40 and 50 (XII) of the Nuclear Law. Two divisions of CNSNS carry out the inspection programme for facilities and activities in Mexico. The inspection of nuclear safety is the responsibility of the Division of Nuclear Safety while the inspection of radiation facilities and activities, radiation protection, emergency preparedness, security and accountability of radioactive sources, physical protection in nuclear installations and safeguards comes under the responsibility of the Division of Radiological Safety. The objective of the CNSNS inspection programme is to ensure that operations related to nuclear safety, radiation protection and nuclear security at facilities and activities are carried out according to established rules and regulations.

In the case of continual, persistent, or extremely serious non-compliance, or a significant release of radioactive material to the environment due to serious malfunctions or damage to a facility or in performance of an activity, CNSNS can take legal action to suspend or revoke the authorization. CNSNS inspectors are empowered to stop operations and to direct necessary actions to restore an adequate level of safety and security.

2.5. STATUS OF RELEVANT REGULATIONS AT THE TIME OF THE INCIDENT

At the time of the incident, regulations on safe and secure transport of radioactive material were not available in Mexico and it was recommended by CNSNS to follow IAEA Safety Standards Series No. SSR-6, Regulations for the Safe Transport of Radioactive Material (2012 Edition). On 10 April 2017, the Regulation for the Safe Transport of Radioactive Material was published in the Official Gazette of the Federation [38]. SSR-6 (2012 Edition) has been superseded by IAEA Safety Standards Series No. SSR-6 (Rev-1) (2018 Edition) [4].

---

[3] The applicant is any person or organization applying to a regulatory body for authorization (or approval) to undertake specified activities [36].
[4] The licensee is the holder of a current licence. The licensee is the person or organization having overall responsibility for a facility or activity [36].

3. THE INCIDENT

At 08:13[5] on 2 December 2013, a worker from Asesores en Radiaciones (ARSA), a company authorized for transport of radioactive material in Mexico notified CNSNS about the theft of a vehicle transporting a teletherapy unit head containing a ^{60}Co radioactive source of an activity of 111 TBq. The vehicle was stolen from a gas station near Tepojaco, in the municipality of Tizayuca, in Hidalgo State.

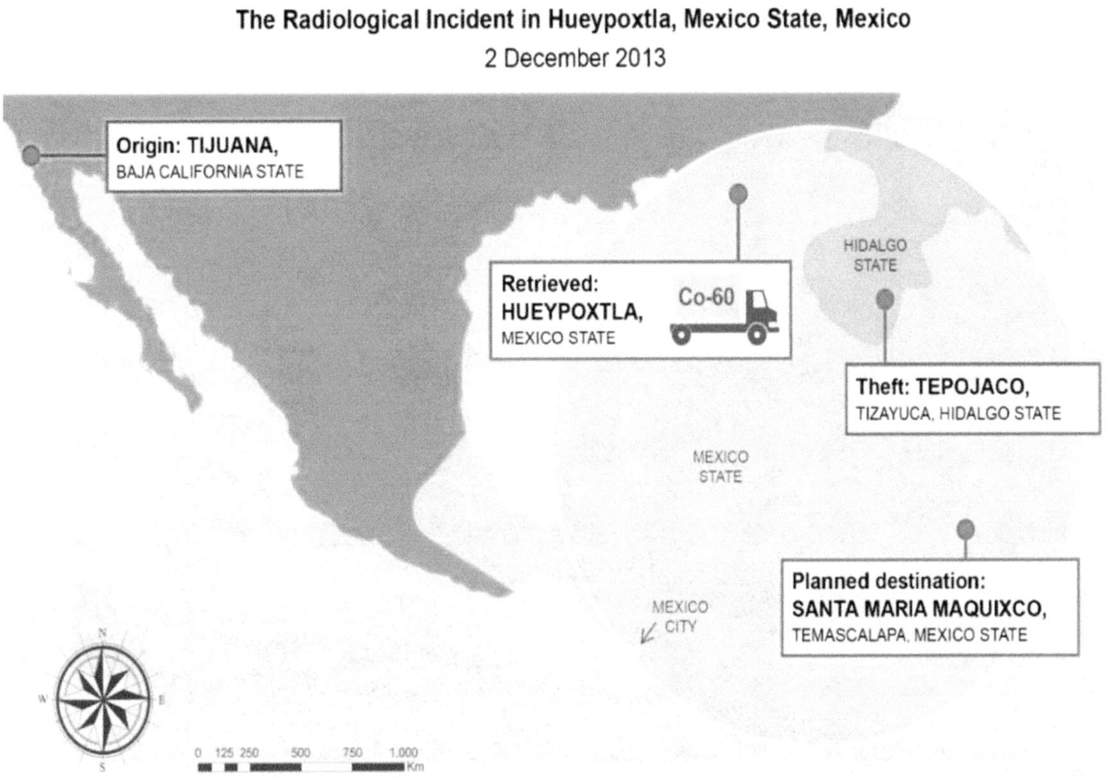

FIG. 2. Locations of the incident in Mexico

The teletherapy unit head belonging to the medical institute "Institute Mexicano del Seguro Social (IMSS)" located in Tijuana, Baja California State was being transported to the radioactive waste storage centre "Centro de Almacenamiento de Desechos Radiactivos (CADER)" located near the town of Santa María Maquixco, Temascalapa municipality, Mexico State.

Following the notification, CNSNS contacted the radioactive transporting company (ARSA) to confirm and authenticate the provided information. CNSNS was informed that at approximately 02:00 on 2 December 2013, the driver and his assistant had been attacked by a group of armed individuals while they were resting at a gas station. The armed individuals snatched the vehicle along with the teletherapy unit head containing the ^{60}Co radioactive source (Fig. 3).

---

[5] All times in this publication are local times in Mexico (UTC-6).

Investigation into the incident revealed that two people were involved in the theft of the vehicle containing the teletherapy unit head. The vehicle along with the teletherapy unit head was sold to a scrap metal dealer on 2 December 2013. Five people working at the scrap metal facility dismantled the teletherapy unit head. During this dismantling process, one person suffered dizziness and started vomiting. In the early morning of 3 December, parts of the teletherapy unit head, along with the ^{60}Co source, were discarded into an uninhabited farming area in the municipality of Hueypoxtla, Mexico State. A local resident (a 40-year-old man, farmworker) saw them discarding parts of the teletherapy unit head in the uninhabited farming area near his house. When they left the location, he visited the site and found some metal pieces, which were lying on the side of the road. He saw a long cylindrical metal piece. He approached and grabbed it. He felt it hot and heavy and lifted it up and placed it on his shoulder and hid it in straws approximately 15 metres away from where it was found.

FIG. 3. Transport of the teletherapy unit head with the ^{60}Co source to the waste disposal facility (picture captured prior to the incident) (courtesy of CNSNS).

4. RESPONSE AT THE NATIONAL LEVEL

This section provides details on the actions performed by various agencies and response organizations in response to the incident. A summary of the source search, localization and recovery activities is presented in Fig. 4.

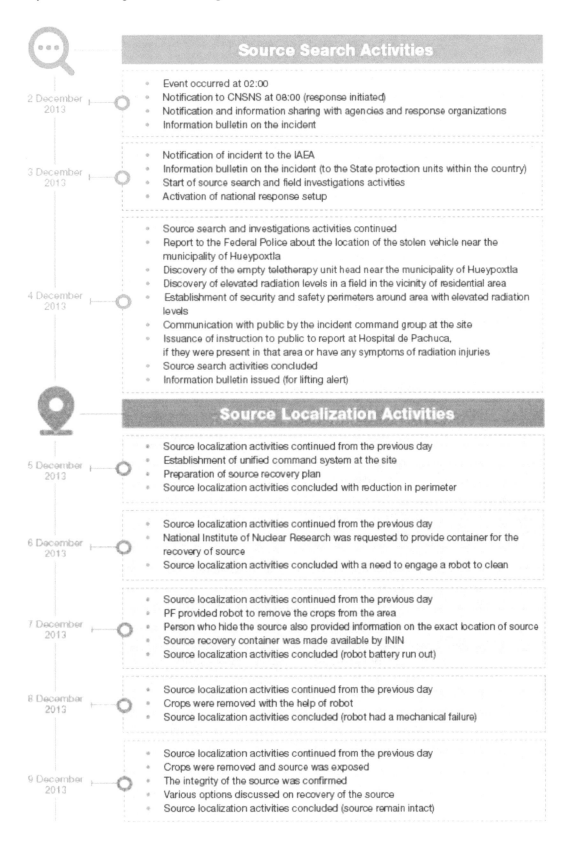

Source Search Activities

2 December 2013
- Event occurred at 02:00
- Notification to CNSNS at 08:00 (response initiated)
- Notification and information sharing with agencies and response organizations
- Information bulletin on the incident

3 December 2013
- Notification of incident to the IAEA
- Information bulletin on the incident (to the State protection units within the country)
- Start of source search and field investigations activities
- Activation of national response setup

4 December 2013
- Source search and investigations activities continued
- Report to the Federal Police about the location of the stolen vehicle near the municipality of Hueypoxtla
- Discovery of the empty teletherapy unit head near the municipality of Hueypoxtla
- Discovery of elevated radiation levels in a field in the vicinity of residential area
- Establishment of security and safety perimeters around area with elevated radiation levels
- Communication with public by the incident command group at the site
- Issuance of instruction to public to report at Hospital de Pachuca, if they were present in that area or have any symptoms of radiation injuries
- Source search activities concluded
- Information bulletin issued (for lifting alert)

Source Localization Activities

5 December 2013
- Source localization activities continued from the previous day
- Establishment of unified command system at the site
- Preparation of source recovery plan
- Source localization activities concluded with reduction in perimeter

6 December 2013
- Source localization activities continued from the previous day
- National Institute of Nuclear Research was requested to provide container for the recovery of source
- Source localization activities concluded with a need to engage a robot to clean

7 December 2013
- Source localization activities continued from the previous day
- PF provided robot to remove the crops from the area
- Person who hide the source also provided information on the exact location of source
- Source recovery container was made available by ININ
- Source localization activities concluded (robot battery run out)

8 December 2013
- Source localization activities continued from the previous day
- Crops were removed with the help of robot
- Source localization activities concluded (robot had a mechanical failure)

9 December 2013
- Source localization activities continued from the previous day
- Crops were removed and source was exposed
- The integrity of the source was confirmed
- Various options discussed on recovery of the source
- Source localization activities concluded (source remain intact)

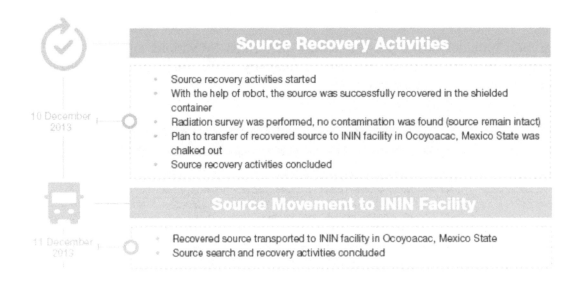

10 December 2013

Source Recovery Activities

- Source recovery activities started
- With the help of robot, the source was successfully recovered in the shielded container
- Radiation survey was performed, no contamination was found (source remain intact)
- Plan to transfer of recovered source to ININ facility in Ocoyoacac, Mexico State was chalked out
- Source recovery activities concluded

11 December 2013

Source Movement to ININ Facility

- Recovered source transported to ININ facility in Ocoyoacac, Mexico State
- Source search and recovery activities concluded

FIG. 4. Source search, localization and recovery activities.

4.1. INITIAL RESPONSE REGARDING SOURCE SEARCH AND RECOVERY

4.1.1. 2 December 2013

Following the authentication and confirmation of the notification regarding the theft of a ^{60}Co radioactive source with an activity of 111 TBq, CNSNS extracted information about the source from the national radioactive source database. CNSNS discovered that as of 3 January 2002, this ^{60}Co source had an activity 458 TBq. The activity of the source was corrected to be 95.24 TBq at the date of the incident.

The incident was reported to the Federal Police "Policía Federal" and to the Committee on International Disarmament, Terrorism and Security "Comité Especializado de Alto Nivel en Materia de Desarme, Terrorismo y Seguridad Internacional (CANDESTI)", an auxiliary body of the national security council that acts as the national authority responsible for coordination at the national level and international liaison.

An information bulletin was prepared by CNSNS for distribution to the relevant government authorities and response organizations by the Civil Protection Agency in Mexico "Coordinacion Nacional de Protección Civil (CNPC)". This information bulletin included details regarding the theft of the radioactive source, potential health hazards that might arise from handling the source unintentionally, recommended actions for responders and the public if they found the radioactive source, contact numbers to contact the relevant authorities if the radioactive source was found. At 13:00 on the same day, the governments of the States of Hidalgo, Veracruz, Puebla, Tlaxcala, Mexico City, Mexico State, Querétaro and San Luis Potosí, as well as the federal authorities were informed through the transmission of this information bulletin.

On receipt of the emergency notification from the Committee on International Disarmament, Terrorism and Security (CANDESTI), the national Security Council "Coordinación de Seguridad Nacional (CoSeNa)" responsible for coordination of national security matters immediately transmitted a message to the Secretariat of Communications and Transports "Secretaría de Comunicaciones y Transports (SCT)", Directorate General of Civil Aviation

"Dirección General de Aeronáutica Civil (DGAC)", Federal Highways and Bridges "Caminos y Puentes Federales (CAPUFE)", General Coordination Centre of the Communications and Transports Secretariat "Coordinación General de Centros SCT (CGCSCT)", Directorate General of Rail and Multimodal Transport "Direccion General de Transporte Ferroviario y Multimodal (DGTFM) and General Coordination of Ports and Merchant Navy "Coordinación General de Puertos y Marina Mercante (CGPMM)" with instructions to implement the necessary measures in accordance with their mandate and authority to locate and secure the source.

The Secretariat of Foreign Relations "Secretaría de Relaciones Exteriores (SRE)" informed the Mexican embassy in Austria and instructed the embassy to maintain close coordination with the IAEA in accordance with Mexico's obligations under the Emergency Conventions [1, 2].

All territorial commands surrounding the state of Hidalgo were instructed to carry out ground investigation, restricting direct contact with the source. The Secretariat of National Defense "Secretaría de la Defensa Nacional (SEDENA)" coordinated with the 37th Military Zone (Santa Lucía, Mexico) for the deployment of a search team to carry out ground investigation on the roads surrounding the town of Tepojaco, in the Tizayuca municipality in the State of Hidalgo and to help the Federal Police to locate the vehicle that was transporting the source.

Through official electronic communications, the Secretariat of Foreign Relations (SRE) reported the event to the embassies of the United States of America, the Netherlands, as well as to the Permanent Missions of Mexico to the international organizations in Geneva, Switzerland and in New York, to ensure that the representatives in the Permanent Missions were informed in a timely manner, in case of possible questions or queries from their counterparts who follow issues relating to nuclear security, disarmament and nuclear non-proliferation.

4.1.2.   3 December 2013

CNSNS notified the IAEA about the incident using a Standard Reporting Form (SRF) through the Unified System for Information Exchange in Incidents and Emergencies (USIE)[6].

A news bulletin with updated information was issued and was disseminated to the Civil Protection States Units "Unidades Estatales de Protección Civil (UEPC)" of the entire Mexican Republic, as well as to the Federal Public Administration "Administración Pública Federal (APF)".

At approximately 13:00 on the same day, CNSNS dispatched four radiation monitoring teams equipped with portable gamma radiation detection equipment to conduct a source search within a 5 km radius of the area in which the theft of the vehicle had occurred. Communication was established with the National Security and Investigation Centre "Centro de Investigación y Seguridad Nacional (CISEN)" for the provision of radiation detection equipment to the Federal Police. During the course of the search, the Federal Police requested a loan of more sensitive detection equipment from CNSNS in order to carry out an aerial survey of the area. CNSNS also requested support from the Federal Police for the transport of the mobile vehicle measuring system of exposure to gamma radiation "Sistema de medición de exposición a la radiación gamma para vehículos móviles (SPARKS)" which was operated by CNSNS. The search ended

---

[6] USIE is a secure IAEA website for designated contact points in IAEA Member States to exchange urgent information during nuclear or radiological incidents and emergencies regardless of whether they arise from accident, negligence or deliberate act.

at approximately 19:00 without any success. CNSNS requested support in radiation monitoring activities from the Laguna Verde Nuclear Power Plant "Central Nucleoeléctrica Laguna Verde (CNLV)", located in the State of Veracruz.

At 18:00, the National Security and Investigation Centre (CISEN) called an extraordinary meeting of the Committee on International Disarmament, Terrorism and Security (CANDESTI). CANDESTI was declared in permanent session in order to promptly follow up on the incident, concentrate the information and communicate it to the relevant response organizations. It was agreed that:

- Each agency would initiate alerting protocols within the framework of its powers and the Secretary of Foreign Relations (SRE) would maintain communication with the Mexican embassy in Vienna, Austria, in case of possible inquiries about the incident.
- The agencies would report any finding to CISEN by event or periodicity and CISEN would ensure collection of information from the agencies and media and proper dissemination among the agencies.
- The Attorney General Office "Procuraduría General de la República (PGR)" would conduct the preliminary inquiry. The agencies would provide PGR with further information and details as become available.
- In terms of public communication, the Civil Protection Agency (CNPC) would inform the media by means of information bulletins about the situation in order to alert and reassure the public.
- CNSNS would inform the IAEA at the same time that a bulletin would be sent to the media.
- In case of localization of the source by any of the agencies, the location of the radioactive source would be communicated to CISEN, the protection protocol would be applied by the National Civil Protection System "Sistema Nacional de Protección Civil (SINAPROC)" at its respective level (federal, State and local), with the support of the Federal Police in order to establish the security perimeter. CISEN would notify PGR and CNSNS for the proper handling of the radioactive source and the Secretariat of Energy "Secretaría de Energia (SENER)" would ensure the radioactive source management (i.e., from the discovery to confinement), regardless of the status of the source.
- Once the source had been located and its status assessed, the National Institute of Nuclear Research "Instituto Nacional de Investigaciones Nucleares (ININ)" would provide arrangement for its disposal.
- The communication scheme would be maintained to give continuity to the incident response actions and the Secretariat of Communications and Transports (SCT) would proceed to investigate the transport company.

4.1.3.　4 December 2013

At 08:00, CNSNS sent two teams equipped with vehicle-based radiation monitoring equipment to perform a search within a 10 km radius of the theft location. The Federal Police searched locations in the municipalities of Tizayuca and Zumpango and the surrounding areas. One of the CNSNS radiation monitoring teams was sent to the Tepojaco gas station where the theft took place to retrieve and evaluate the information recorded by the CCTV cameras. The Federal Police continued with the deployment of personnel in the towns near Tizayuca and Zumpango.

At 12:40, the Federal Police from the mixed operations base station (SEDENA, State Police and Municipal Police) received information from SEDENA that a vehicle, similar to the one that had been transporting the source, was located near the Hueypoxtla municipality. The Federal Police officials were sent to verify the information and to search the area for the source.

The Federal Police officials were approached by a local resident (a 40-year-old farmworker) who informed them about a suspicious metallic object in the backyard of his house. Upon investigation, the Federal Police identified the empty teletherapy unit head (Fig. 5).

The Federal Police detected background radiation levels in the vicinity of the house. Elevated radiation levels were detected in an uninhabited farming area approximately one kilometer east of the house.

At 14:30, the Federal Police reported their findings to CNSNS. CNSNS advised the Federal Police to perform radiation measurements near the empty radioactive source shielding and identify any serial number on the device that could provide information about the radioactive source. CNSNS instructed its radiation monitoring team to visit the location identified by the Federal Police.

FIG. 5. The empty radioactive source shielding (courtesy of CNSNS).

On arrival at the site, the Federal Police briefed the CNSNS team about the radiation measurements they had conducted and the discovery of the empty teletherapy unit head. The Federal Police guided the CNSNS team to the areas where elevated radiation levels had been detected.

Based on the radiation level measurements, CNSNS advised the Federal Police to establish a security perimeter of approximately 1 km radius around the uninhabited farming area to prevent access to the area by members of the public and non-essential personnel.

The CNSNS team, equipped with a telescopic radiation detector, observed substantial increase in the radiation levels in the uninhabited farming area. CNSNS distributed thermoluminescent dosimeters (TLDs) to the responders working in the area. Based on the results of the radiation survey, CNSNS established a perimeter within which the gamma dose rates were greater than 100 μSv/h. Later, another perimeter was established where the gamma dose rates were greater than 500 mSv/h.

At 22:30 due to the lack of light, the search activities were stopped. The Federal Police and the army were asked to secure and guard the area to ensure that only authorized personnel could enter the cordoned-off area.

With the discovery of the empty teletherapy unit head and the elevated radiation levels from the ^{60}Co source (hidden in the uninhabited farming area), a last information bulletin was issued notifying the lifting of the alert by CNPC.

4.1.4. 5 December 2013

The teams of the Federal Police, the Secretariat of the Navy "Secretaría de Marina Armada de México (SEMAR-AM)", Laguna Verde Nuclear Power Plant (CNLV) and CNSNS arrived at the scene. The local civil protection agency highlighted the need to establish a command system at the scene. The unified command system composed of personnel from CNSNS as incident commander and radiological advisor, CNLV, SEMAR-AM as response and support groups for recovery tasks and the Federal Police as security forces was established. The radiation survey was continued to be performed to delineate areas with elevated radiation levels aiming to reduce the search area for localization and recovery of the radioactive source. At CNSNS headquarters, heads of CNSNS, CNLV and SEMAR-AM started drawing up an action plan for the recovery of the source.

4.1.5. 6 December 2013

CNLV and SEMAR-AM teams arrived at Hueypoxtla and reinforced the CNSNS team. The approximate location of the source was determined by the CNLV team as the radioactive source was covered in straws. For verification and subsequent recovery of the radioactive source, there was a need to remove the straws under which the source was hidden. Further, for the safe recovery and transport of the recovered source, National Institute of Nuclear Research (ININ) was requested to provide a suitable transport container. For the intended purpose, such a container was not readily available at the ININ facilities, and some modifications were made to an existing container to contain the radioactive source.

4.1.6. 7 December 2013

The teams from CNSNS, CNLV, SEMAR-AM and the Federal Police started planning to remove crops from the area with a robot belonging to the Federal Police to enable the source to be precisely located. On investigation by the Federal Police, the resident of the house from where the empty container was discovered was willing to indicate where the source was hidden in the uninhabited farming area. With the help of this person, the exact location of the (unshielded) source was determined. CNLV and CNSNS staff asked the person about the amount of time he had spent near the source and asked him to show them his shoulder, being able to observe swelling in the left trapezius of the individual. Medical examination was offered, but the person declined. The transport container was made available by ININ.

4.1.7.   8 December 2013

The teams of CNSNS, CNLV, SEMAR-AM and the Federal Police returned to the uninhabited farming area to continue the crop removal process using the robot. The crop removal process was stopped due to a mechanical failure in the robot and the robot was removed from the area. SEMAR-AM brought concrete containers and lead blankets from the CNLV facilities in Veracruz, to provide shielding from the source for the recovery operations. However, it was determined that this material was not suitable for the recovery operations.

The resident of the house from where the empty container was discovered, reported to CNSNS that he had seen individuals discarding parts of the teletherapy unit head in the uninhabited farming area near his house in the early morning of 3 December. After they had moved away, he had approached the site and discovered some metal pieces, picked up the source and thrown it away approximately 15 meters inside the straws, in the uninhabited forming area.

4.1.8.   9 December 2013

The CNLV team finished the crop removal process and uncovered the source which was hidden in the crops (Fig. 6). It was confirmed that the source was not damaged. Plans for the recovery of the source were discussed in a meeting attended by personnel from SENER, ININ, CNSNS and CNLV. The following options were discussed:

- Removing the source with the robot and depositing it in the shielding provided by ININ.
- Removing the source using a rope (knot and tie-down technique) and depositing it in the shielding provided by ININ.
- Removing the source with a crane and depositing it in the shielding provided by ININ.

FIG. 6. Exposed radioactive source (courtesy of Federal Commission for Electricity of Mexico).

4.1.9.   10 December 2013

After the robot was repaired, the recovery operation continued. To contain the recovered radioactive source, the modified container was brought by ININ to the site. Images of the source were taken by the robot camera, and it was confirmed that the source was intact. Using the robot, the source recovery operation was initiated and after two attempts, the source was successfully placed in the container (Fig. 7). After placing the recovered source inside the container, the lid of the container was placed and the radiation levels on the surface of the container were measured by the CNSNS team. The radiation levels at the surface of the container were below the safe levels recommended in the IAEA safety standards [4].

After successful recovery of the source into the shielded container, a radiation survey in the uninhabited farming area was performed and only background radiation levels were detected, confirming that the source remained intact, and no contamination occurred in the uninhabited farming area.

Arrangements were made for the transport of the source to the ININ facilities in Ocoyoacac, Mexico State. The time, route and escort arrangements were agreed between CNSNS, ININ, the Federal Police and the transport service provider.

FIG. 7. The container with the recovered radioactive source (courtesy of CNSNS).

During the source recovery, response actions were taken to remove the straws that covered the source by remote means, in order to avoid unnecessary radiation exposure of the responders. CNSNS stated that the 'ALARA' ('as low as reasonably achievable') principle was not fully implemented by some of the participating institutions which concentrated on the urgency of immediate removal of the source without taking into account the radiation exposure of the responders.

4.1.10.   11 December 2013

CNSNS instructed ININ to receive the ^{60}Co source at its facility in Ocoyoacac, Mexico State. The CNSNS personnel verified the measures adopted in terms of nuclear security. CNSNS supervised the transport of the source to ININ facility in Ocoyoacac, Mexico State.

4.2. PUBLIC COMMUNICATIONS

The incident command group which consisted of representatives from CNSNS and the Ministry of Health informed the public on 4 December 2013 on the potential danger in handling and being close to the radioactive source. The CNSNS representative instructed the members of the public, who might have been in contact with, or in the immediate vicinity of the source, to report to the Hospital de Pachuca for the estimation of the radiation doses received and for the identification of any need for medical follow-up. The members of the public from Hueypoxtla village raised concerns and several inquiries on the radiological situation at the site and the measures being taken for the safe recovery of the radioactive source. The CNSNS representative answered these inquiries. With several concerns of the locals on the situation and the ongoing activities regarding the recovery of the source, the interaction with the public was discontinued as the situation become unstable. The representatives of the incident command group were safely removed from the crowd by the Federal Police.

All relevant agencies were instructed to direct requests for information or interviews from national and/or international media to the Secretary of Foreign Relations (SRE). SRE was made responsible to transmit any request of information to the corresponding authorities, bearing in mind that only two organizations were authorized to provide official information on the incident (i.e. SENAR and CNSNS) to address technical issues and the spokesperson of the Presidency of the Republic for issues corresponding to politics.

4.3. MEDICAL MANAGEMENT ACTIVITIES

On 3 December 2013, the National Centre for Preventive Programmes and Disease Control "Centro Nacional de Programas Preventivos y Control de Enfermedades (CENAPRECE)" convened an extraordinary meeting of the subcommittee on emerging diseases to report on the situation and direct the response for the safety of the population potentially at risk. Representatives from the Directorate General of Epidemiology "Dirección General Adjunta de Epidemiologia (DGAE)", Institute for Diagnose and Reference "Instituto de Diagnóstico y Referencia (InDRE)", General Directorate of Health Promotion, Federal Commission for the Protection of Sanitary Risks "Comisión Federal para la Protección contra Riesgos Sanitarios (COFEPRIS)", IMSS, Institute of Security and Social Services for State Workers "Instituto de Seguridad y Servicios Sociales de los Trabajadores del Estado (ISSSTE)", SEDENA and SEMAR-AM participated in the meeting. By internal agreement of the subcommittee, it was decided to issue an epidemiological emergency declaration for the states of Hidalgo, Mexico and the federal district.

On 4 December 2013, when the ^{60}Co source was located in Hueypoxlta, Mexico State, the operations centre was established in the municipal Hospital 'Hermenegildo Galeana'. A working team made up of personnel from the Directorate of Epidemiological Emergencies and Disasters, Veracruz Health Services and National Institute of Medical Sciences and Nutrition 'Salvador Zubirán', "Instituto Nacional de Ciencias Médicas y Nutrición 'Salvador Zubirán' (INCMNSZ)" was established. The team moved to the State of Hidalgo, where meetings were held with officials from the State's health services, in order to exchange information and establish a coordinated and integrated approach on the medical management. It was agreed that people who might have been exposed to radiation should be transferred or should report to the Hospital de Pachuca for assessment of radiation doses and medical follow up.

The response to the incident was managed within the framework of the subcommittee on emerging diseases, in coordination with the participating institutions and the health services of the States of Hidalgo and Mexico. In coordination with National Institute of Medical Sciences and Nutrition 'Salvador Zubirán' (INCMNSZ) and health services of the States of Hidalgo and Mexico, medical treatment was provided, as necessary, to all persons suspected of radiation exposure due to the radioactive source.

On 8 December 2013, CNSNS coordinated with Veracruz State's Ministry of Health for providing necessary support in medical examination of the individuals who might have been exposed to the radioactive source and might have received radiation doses. Subsequently, the Veracruz State's Ministry of Health coordinated with the Federal Ministry of Health for the support in medical examinations, as it became necessary. The activation of staff by the Federal Ministry of Health was confirmed on 9 December along with that of Veracruz State's Ministry of Health.

On 9 December 2013, individuals who might have been exposed to the source were examined at the Hospital de Pachuca by the representatives of Veracruz State and the federal ministries of health accompanied by CNSNS personnel. The CNSNS team visited Hueypoxtla for the medical examination of the individual who assisted in locating the source and of the individual who was believed to have been in contact with the source, while it was still shielded. It was identified that the individual who assisted in locating the source had radiation injuries on the left shoulder and the right leg. This individual was moved to the Hospital de Nutrición in Mexico City for treatment and follow-up. The second individual was also examined, and it was identified that this individual had not received a significant radiation dose and no symptoms of radiation exposure were found.

4.4. RADIATION DOSES RECEIVED BY THE PUBLIC AND EMERGENCY WORKERS

To identify the individuals who might have been exposed due to the radioactive source, a field investigation was conducted by the federal ministry on 10 December 2013. The members of the public who visited or passed by the uninhabited farming area were questioned about their location and time that they had spent where the source was hidden. Event reconstruction was performed to estimate the radiation doses received by the members of the public. Based on this field investigation, a total of fifty-nine individuals were identified as possibly exposed in this incident. The radiation doses for each individual were estimated based on their distance from the source location and the time which they spent near the source. Based on this study, it was identified that thirty-seven out of fifty-nine individuals had not been present at the relevant dates and times and were excluded from further studies. For the remaining twenty-two individuals, estimation of the received doses and assessment of the acute radiation exposure was performed based on reconstruction of events and on evaluating their possible exposure.

On 13 December 2013, for ten individuals out of these twenty-two individuals, a request was made to ININ by CNSNS and the Federal Ministry of Health to perform biological dosimetry studies. These ten individuals also included four persons that presented symptoms that could be associated with the acute radiation syndrome.

The biological dosimetry studies of the ten individuals were performed on 15 December 2013 by ININ. It was identified that only one person had received a dose in excess of the limit specified in the Mexican regulations to prevent non-stochastic effects among occupationally exposed personnel (500 mSv annual whole body effective dose)[7]. This individual, as mentioned in section 4.1.3, helped the Mexican authorities to locate the source.

The Ministry of Health informed CNSNS about a second group of five persons, who had been allegedly involved in the dismantling of the teletherapy unit head and who had been exposed to the radioactive source. On request of CNSNS, ININ collected blood samples and conducted biological dosimetry studies of this second group of people. One person was identified as having received a dose in excess of the limit specified in the Mexican regulation to prevent non-stochastic effects among occupationally exposed personnel. In total, dosimetry studies were performed for twenty-seven individuals as presented in Fig. 8.

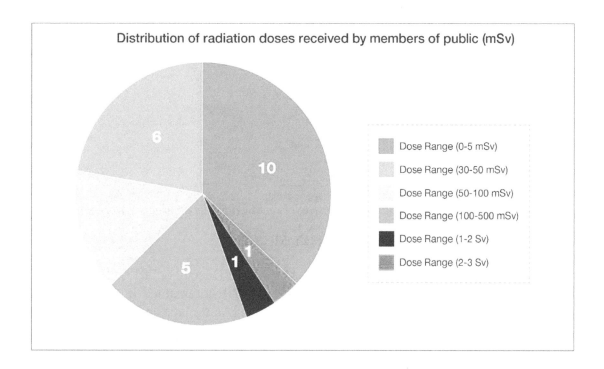

FIG. 8. Distribution of radiation doses received by members of public.

For the workers involved in the recovery process, a limit of 50 mSv effective dose was set. The average dose received by the emergency workers was less than 3 mSv, and the highest value was approximately 20 mSv.Figure 9 provides the details of the radiation doses received by the emergency workers in the source search and recovery operations.

---

[7] Since in the Mexican regulation there are no exposure limits to the public in case of a radiological emergency, it was agreed to use the limit of non-stochastic effects for the occupationally exposed personnel.

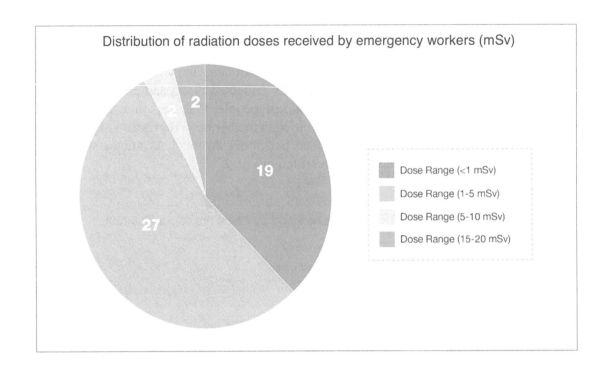

FIG. 9. Distribution of radiation doses (mSv) received by the emergency workers.

4.5. INTERNATIONAL COORDINATION BY MEXICO

In compliance with the international conventions [1, 2], on 3 December 2013, CNSNS notified the IAEA of the event through the USIE website. The IAEA remained in contact with the Permanent Mission of Mexico in Vienna and CNSNS. The IAEA made an offer of good offices to Mexico when the source was lost, in addition to standing-by to provide assistance, if requested.

In addition, the USA National Nuclear Safety Agency (NNSA), Department of Energy (DOE), submitted a request for information on the event and the progress that had been made regarding the search and recovery of the radioactive material. An offer of assistance in source search and recovery was also extended to Mexico by the USA.

4.6. INVESTIGATIONS AND FIXATION OF LIABILITIES

In order to address irregularities in the authorization of issuance of permit[8] and licence[9] for the transport of radioactive material, the SCT, CNSNS, SENER and the Federal Police were summoned at the proposal of the National Security Coordination. Four meetings were held in 2014 and the following were agreed:

- Standardization of the process of issuing permits and licences for the transport of radioactive material needs to be based on legal concurrence;
- Exchange of information needs to occur in order to detect possible irregularities;

---

[8] Transport permit issued by SCT to the transport company
[9] License issued by CNSNS for the transport of radioactive material

- Adaption of legal and regulatory framework that establishes radiation safety and security measures needs to be performed.

4.6.1. Judicial investigations and prosecutions

On 4 December 2013, CNSNS notified CANDESTI that the stolen device had been located, without the ^{60}Co source, and that the radioactive source was presumably located in an uninhabited farming area, approximately one km from the place where the device was found. A legal statement was obtained from the people involved in the discovery of the device, who stated that they had found the piece abandoned near their home, so they decided to take it and sell it as scrap metal.

On 5 December 2013, legal investigation personnel went to the place where the device was found, in order to carry out the legal inspection without entering the cordoned area. Forensic experts intervened in order to search for genetic samples and fingerprint samples of the people who might have manipulated the device. According to the radiological advisors, manipulation of the device no longer represented any risk as the device parts consisted of source shielding material without any contamination.

On 6 December 2013, information was received from Hidalgo State's police, in which they reported the arrest of five persons who had been exposed to the radioactive source, while working at their scrap metal facility. They were transferred to the Hospital de Pachuca in order to receive medical attention and were later placed at the disposal of the Agent of the Public Ministry of the Federation "Agente del Ministerio Público de la Federación (AMPF)". On the same day, a 'present and locate order' was issued against those who had stolen the vehicle with the radioactive source. In their ministerial statement, it was pointed out that a person had gone to the scrap metal facility of one of the detainees, to sell the vehicle, including the teletherapy unit head. At the facility, they had dismantled the teletherapy unit head. One person (16 year old man) had suffered from dizziness and started vomiting. They had thrown away parts of the teletherapy unit head along with the ^{60}Co source in the uninhabited farming area in the municipality of Hueypoxtla, Mexico State. On 7 December 2013, based on available information, a portrait of a person involved in the theft was made. Later on, two persons were also arrested for being involved in the theft of the vehicle and of the radioactive source.

On 1 February 2014, imposition of administrative and criminal sanctions of preliminary inquiry was filed by the Attorney General Office (PGR), bringing criminal charges against seven detainees for their probable responsibility in committing the following crimes:

- Organized crime of vehicle theft;
- Theft of vehicle;
- Dismantling of a stolen motor vehicle;
- Reception of a stolen motor vehicle;
- Theft of teletherapy unit head and ^{60}Co source;
- Abandonment of radioactive substances.

4.6.2. CNSNS investigations and regulatory enforcement actions

As a consequence of the incident, on 14 February 2014, CNSNS initiated the administrative procedure for imposing sanctions on ARSA, the company responsible for the transport of radioactive material under the following charges which were considered violations of the safety regulations and licence provisions during the transport of radioactive material:

- No security was provided to the vehicle that transported the radioactive source;
- The vehicle that was used for the transport of the radioactive material was parked at gas stations where the driver and his assistant made an overnight stop, which is contrary to the provisions of the conditions indicated in the authorization issued by CNSNS for the transport of radioactive material;
- The placards were not visibly displayed on both the sides of the cargo compartment of the vehicle that was used for the transport of the radioactive source. The radiation labels were hidden by the covering blanket and the vehicle was not labelled with radiation placards which is contrary to the provisions of the conditions established in the company's authorization for the transport of radioactive material;
- The radiation labels were not placed on the container, as required by the General Radiological Safety Regulations and in accordance with the conditions established in the authorization for the transport of radioactive material;
- The vehicle that was transporting the referred container was driven by a person and his assistant, who were not registered as occupationally exposed workers by CNSNS in the authorization for the transport of radioactive material;
- The vehicle that was used for the transport of the package, was not authorized by CNSNS to carry out transport of radioactive material in accordance with the company's authorization for the transport of radioactive material;
- The transport of the radioactive source was carried out without the presence of at least two persons authorized as occupationally exposed workers, as established in the company's authorization for the transport of radioactive material;
- The radiation levels were measured around the package and the vehicle only prior to the initial departure of the vehicle, which contravenes the condition set forth in the company's authorization for the transport of radioactive material, that the measurements need to be carried out after each stop-over.

From the breaches of the provisions indicated in items 1–4 of the above list, it was concluded by CNSNS that, if these had been fulfilled during the transport of the radioactive source, it was highly probable that the theft of the vehicle and the dismantling of the teletherapy device would have been avoided. Subsequently, the following would also have been avoided:

- The unnecessary exposure of the public, including that of the person who found it and who suffered significant injuries due to the exposure to radiation, as well as various people who passed near the place where the source was found outside its container;
- The radiation dose that the CNSNS personnel and the first responders (Federal Police, SEMAR-AM, ININ, CNLV and others) received as a result of the emergency response and the recovery of the source;
- The panic, alarm and social conflict that was generated, both in the place where the ^{60}Co source was found outside its container (Hueypoxtla, Mexico State) and in the surrounding areas;
- The high economic costs that the federal government and the local government incurred in order to respond to the emergency and to mitigate its consequences.

Derived from the establishment of the administrative sanction procedure and taking into account the evidence, arguments and allegations made by the company, the resolution was issued on 8 May 2014, by which the following were finally determined:

- The definitive cancellation of the licence for the transport of radioactive material held by the company ARSA;
- The application of a penalty for a total amount of Mex$ 601 524.

4.6.3. Investigations and imposition of fines by the Secretariat of Communications and Transports

Two inspections to the permit holder were carried out by the Secretariat of Communications and Transports (SCT). The first inspection of the general cargo was performed in 3–6 December 2013. Irregularities were determined resulting in a sanction consisting of the payment of a fine of Mex$ 229 898. The second inspection of the specialized cargo was performed in 4–6 December 2013. Irregularities were determined resulting in a sanction consisting of the payment of a fine of Mex$ 19 428.

5. INTERNATIONAL RESPONSE – RESPONSE FROM THE IAEA

On 3 December 2013, the IAEA's IEC received a notification from Mexico that a vehicle transporting a teletherapy unit head, containing a ^{60}Co source with approximate activity of 111 TBq was stolen. The CNSNS confirmed that the Mexican law enforcement authorities and civil protection agency had been alerted. A squad composed of police and personnel from the regulatory authority has been dispatched to the towns near the Tepojaco area to begin searching for the missing source, and a statement has been released to the press to notify the public of this event.

On receipt of the initial information from Mexico, the Incident and Emergency Response System (IES) of the IAEA was activated [39]. In accordance with the IAEA response protocols, the information was authenticated with the Mexican Point of Contact (PoC) and was published on USIE. On the same day, the IAEA sent an 'offer of good offices' to Mexico for any assistance it might need in responding to the incident through the IAEA Response and Assistance Network (RANET) [40].

Subsequent to the initial notification on 3 December 2013, the CNSNS information bulletin along with the images of the source and the transport vehicle were published on USIE.

On 4 December 2013, updates on response activities were shared with the IAEA. The IAEA was informed that radiation monitoring teams had been deployed to monitor the areas of Tecamac, Ojo de Agua, Ozumbilla, Heroes de Tecamac, Jardines de Morelos, Las Palomas, San Tequisquiac, San Juan Zicaltepec, La Planada, Coyotepec, Apaxco, Huehuetoca, Santa Maria Apaxco, Tepojaco, Camino Lechería Texcoco, Tepexpan, Chinconcuac, Avenida Central, Avenida Rio de los Remedios, Avenida R1, Central de Abastos de Ecatepec, Las Americas, Avenida Primero de Mayo, Via Morelos, Cerro Gordo, Xalostoc, Alta Villa, Autopista Mexico Pachuca, San Juanico and San Cristobal Ecatepec. No trace of the truck or the source had been found. The public and the media had been notified of the incident, the potential implications and the precautions to be taken. Telephone numbers were made available to the public to report the source and/or the truck in case a member of the public had any relevant information. The IAEA's IEC remained continuously engaged with the Mexican counterparts to closely monitor the developing situation.

On 5 December 2013, the IEC shared an update on the incident with all the emergency contact points on USIE. On the same day, the IAEA issued a press release regarding the localization of the missing source stating that "the IAEA remain in close contact with the Mexican authorities. It believes the actions taken in response to the discovery of the source are appropriate and follow Agency guidance for this type of event" (see the Appendix).

On 6 December 2013, further information on the recovery of the discovered source was shared with the IAEA through USIE. Mexico had noted that the radioactive source that had been missing had been located. Law enforcement had tracked the radioactive source to an uninhabited farming area near the town of Hueypoxtla in Mexico State, very close to where the truck had been stolen, at around 14:00 on 4 December 2013. The radioactive ^{60}Co source contained in the teletherapy unit head had been removed from its protective shielding, but there was no indication that it had been damaged or broken up and there was no sign of contamination in the area. Police had secured the area around the source to a distance of 500 m.

Radiation measurements in a village close to where teletherapy unit head had been found were recorded to be at background levels. Persons that potentially came close to the source had been monitored and no contamination had been found. The recovery operation had been planned and implemented.

On 12 December 2013, subsequent to the recovery of the radioactive source, the IAEA issued a second press release regarding the safe recovery of the radioactive source. Along with sharing response activities, it was highlighted that "based on the information available, the Mexican authorities and the IAEA believe the general public is safe and will remain safe" (see the Appendix). The two IAEA press releases are provided in the Appendix.

On 10 August 2015, an INES rating form was published by the Mexican National INES officer on USIE. The incident was categorized at level 3 based on criteria provided in the INES user's Manual 2008 Edition [41].

6. OBSERVATIONS, LESSONS IDENTIFIED AND GOOD PRACTICES

This section presents observations, lessons identified, and good practices in the response to the incident in Hueypoxtla by national authorities and response organizations.

6.1. OBSERVATIONS AND LESSONS IDENTIFIED

6.1.1. Implementation of regulatory requirements

Observation: Although CNSNS regulations on radiation safety establish general requirements on transport safety, at the time of incident, there was no national regulation on the safe transport of radioactive material, and it was recommended by CNSNS to SSR-6 (2012 Edition).

Lesson identified: Subsequent to the incident, work on the development of national (Mexican) regulations on the safe transport of radioactive material was initiated and these were promulgated on 10 April 2017.

6.1.2. Development of a national radiological emergency response plan

Observation: At the time of the incident, a national radiological emergency response plan and relevant procedures were not available, and an ad-hoc arrangement was made for coordinating and performing response actions by the relevant agencies and response organizations at the national level.

Lesson identified: A national radiological emergency response plan needs to be in place that integrates all relevant plans for emergency response in a coordinated and integrated manner and describe the roles and responsibilities of relevant agencies and response organizations.

6.1.3. Emergency management system

Observation: The lack of an established emergency management system was observed in response to the incident and ad-hoc arrangements were made to delegate and discharge responsibilities for the source search and recovery activities.

Lesson identified: An integrated and coordinated emergency management system needs to be established, documented and maintained.

6.1.4. Allocation of roles and responsibilities

Observation: The roles and responsivities carried out by the national authorities and response organizations were not allocated and documented in the respective plans and procedures.

Lesson identified: Roles and responsibilities for preparedness and response for a nuclear or radiological emergency need to be clearly allocated in advance among authorities and response organizations at the local and national level and documented in the national radiological emergency response plan.

6.1.5. Coordination between response organizations

Observation: Although, the authorities and response organizations coordinated timely and effectively in response to this incident without having a national radiological emergency response plan, it was observed that there is a need to establish and document coordination arrangements between authorities and response organizations at the national level.

Lesson identified: Arrangements for the coordination between authorities and response organizations, including security and law enforcement agencies, need to be established.

6.1.6. Unified command and control system for emergency response

Observation: There was no unified command and control system for managing the source search and recovery activities at the site and ad-hoc arrangements were made to establish the incident command group.

Lesson identified: Arrangements for a clearly specified and unified command and control system for emergency response as part of the emergency management system needs to be established and documented.

6.1.7. Notification and information sharing

Observation: There were no pre-established points of contact for notification and information sharing in each response organization. In response to this incident, notification and sharing of information was made between authorities and response organizations through personnel identified in relevant training events and through the coordination points of contact established during the Pan American Games held in Guadalajara in 2011.

Lesson identified: There is a need to establish notification and information sharing arrangements to notify the appropriate organizations and to provide sufficient information for an effective response and initiate a coordinated and pre-planned response upon notification of an incident or event.

6.1.8. Instructions to the Public

Observation: Public information bulletins in response to this incident were issued on ad-hoc basis. No arrangements on public information and to issue instructions and warnings in the case of such an incident were available.

Lesson identified: Arrangements to provide the public with information and instructions in order to identify and locate people who might have been affected and who might need to take response actions such as decontamination, medical examination or health screening need to be established.

6.1.9. Managing the medical response

Observation: In response to this incident, it was noted that the medical professionals were not educated and equipped for the medical treatment of individuals exposed to radiation.

Lesson identified: Medical professionals need to be made aware of the clinical symptoms of radiation exposure, and of the appropriate notification procedures and other emergency response actions to be taken if a nuclear or radiological emergency arises or is suspected.

6.1.10. Public communication

Observation: It was noted that designated spokespersons of each response organization were not respected and abided by all agencies and response organizations involved in the response to the incident. Further, in responding to queries of the public during the press release, the situation showed signs of becoming unstable and the Federal police discontinued this interaction by removing the representative of the incident command group from among the crowd.

Lesson identified: Arrangements need to be made to ensure that information provided to the public by authorities and response organizations in such incidents is coordinated and consistent, with due consideration for the protection of sensitive information in a nuclear security event. Further, public trust on the authorities and response organizations needs to be developed in the preparedness phase.

6.1.11. Authorities for emergency preparedness and response

Observation: At the incident location, some people who were not involved in response activities were found inside the safety perimeter, despite CNSNS instructions. In addition, the ALARA principle was not fully implemented for all responders due to lack of training of the personnel and due political pressure to urgently complete the recovery activities.

Lesson Identified: Authorities for preparedness and response for a nuclear or radiological incident need to be clearly established. Conflicting or potentially conflicting and overlapping roles and responsibilities need to be identified and conflicts need to be resolved at the preparedness stage through the national coordinating mechanism. Political pressures should not take precedence over the emergency response.

6.1.12. Analysing the emergency and emergency response

Observation: To address irregularities in issuance of permits and licences for the transport of radioactive material, some meetings between authorities and response organizations were held after this incident, however a need was identified to have arrangements in place to convene feedback meetings to analyse the emergency and emergency response.

Lesson identified: The incident and the incident response need to be analysed to identify actions to be taken to avoid the occurrence of similar incidents and to improve emergency arrangements.

6.1.13. Training, drills and exercises

Observation: In response to this incident, a need was recognized for training of first responders and medical professional in basic radiological protection and in managing the medical response.

Lesson identified: Arrangements need to be established for emergency response personnel to take part in regular training, drills and exercises to ensure that they can perform their assigned response functions effectively in response to such incidents.

6.2. GOOD PRACTICES IDENTIFIED

6.2.1. Coordination between response organizations

Even though there was no emergency plan for radiation emergencies, some of the responders had trained and worked together during the Pan American Games Guadalajara 2011 in the nuclear security arrangements of major public events. When the incident occurred, they knew which Agencies could help tackling it and the right personnel to contact.

6.2.2. Providing information to the public

When the incident occurred, information was shared with the public about the event and the risks involved. Consequently, the people who were involved with the theft of the vehicle and the teletherapy unit head became aware of the associated risks and they immediately disposed of the truck and device. Also, when a member of the public identified the shielding (teletherapy unit head), they contacted the authorities which helped in locating and recovering the radioactive source.

6.2.3. Sharing resources

It is almost impossible that one agency has all the resources to respond to this type of incident on its own. In this incident, the following agencies and organizations responded as follows:

- The regulatory authority, the nuclear power plant and the nuclear research institute shared radiation detection and radiation dosimetry equipment with the responders.
- The navy shared power sources and a helicopter.
- The Federal Police provided their bomb dismantling robot and armoured cars that were used as shielding for the responders.

7. CONCLUSIONS

The Hueypoxtla incident illustrates a radiological incident that occurred during the transport of a radioactive source. Further, a security event that might not be directly related to the radioactive source itself might result in a radiological emergency at an unforeseen location within a State. This incident highlights the need for having emergency preparedness and response arrangements for a nuclear or radiological emergency at an unforeseen location. In response to this incident, it was concluded that such an emergency is beyond the capabilities of a single agency and there is a need to have a national radiological emergency response plan which describes the roles, responsibilities and resources of relevant agencies and response organizations.

The years following this incident have shown to Mexican authorities that most radiological events in Mexico occur due to security issues. Consequently, Mexico established national security requirements to prevent malicious acts in the future. It was also concluded that not only is it necessary for the emergency response agencies to be adequately trained but there is also a need to train the security and law enforcement agencies on the basics of radiological protection and in the response to radiological emergencies. In this regard, special groups have been established inside these agencies with the specialized skills and equipment to ensure appropriate response and management of such type of event.

Appendix

IAEA PRESS RELEASES REGARDING THE INCIDENT IN HUEYPOXTLA

A.1.   FIRST IAEA PRESS RELEASE

5 December 2013

Mexico Says Stolen Radioactive Source Found in Field

FIG. 10. The international radiation symbol, or trefoil, indicates hazardous radioactive material. (Graphic: IAEA)

Mexico has informed the IAEA's Incident and Emergency Centre (IEC) that it has located a dangerous radioactive source that had been missing since the truck on which it was being transported was stolen on 2 December 2013.

Mexico's "Comisión Nacional de Seguridad Nuclear y Salvaguardias (CNSNS)" said law enforcement authorities tracked the teletherapy device down to a field near the town of Hueypoxtla in Mexico State, very close to where the truck was stolen, at around 14:00 (20:00 UTC) on 4 December 2013.

The radioactive cobalt-60 source contained in the device has been removed from its protective shielding, but there is no indication that it has been damaged or broken up and no sign of contamination to the area. Police have secured the area around the source to a distance of 500 metres.

The source, with an activity of 3000 curies (111 terabequerels), is considered Category 1. The IAEA defines a Category 1 source as extremely dangerous to the person. If not safely managed or securely protected, it would be likely to cause permanent injury to a person who handled it or who was otherwise in contact with it for more than a few minutes. It would probably be fatal to be close to this amount of unshielded radioactive material for a period in the range of a few minutes to an hour.

Mexican authorities are assessing potential radiation exposure to persons who may have been close to the unshielded source, and hospitals have been alerted to watch for symptoms of such exposure.

People exposed to the source do not represent a contamination risk to others. Based on the information available, the Mexican authorities and the IAEA believe the general public is safe and will remain safe.

CNSNS and the Instituto Nacional de Investigaciones Nucleares (ININ) are preparing plans to recover and secure the source.

The IAEA remains in close contact with the Mexican authorities. It believes the actions taken in response to the discovery of the source are appropriate and follow Agency guidance for this type of event.

A.2.   SECOND IAEA PRESS RELEASE

Mexico Says Stolen Radioactive Source Found in Field

FIG. 11. The international radiation symbol, or trefoil, indicates hazardous radioactive material. (Graphic: IAEA)

Mexico has told the IAEA's Incident and Emergency Centre (IEC) that it has safely recovered the dangerous radioactive source that had been abandoned in a field after being stolen last week.

Mexico's nuclear regulator, the "Comisión Nacional de Seguridad Nuclear y Salvaguardias (CNSNS)", said the delicate and complex recovery operation was successfully completed on the evening of 10 December 2013 using a Federal Police robot.

The highly radioactive source had been removed from its protective shielding but is intact and undamaged. There is no contamination to the surrounding area.

The Category 1 cobalt-60 teletherapy source, which was formerly used for cancer treatment, had been on a truck that was stolen on 2 December 2013. The source was located two days later, abandoned among crops in a field near the town of Hueypoxtla in Mexico State.

The IAEA defines a Category 1 source as extremely dangerous to the person. If not safely managed or securely protected, it would be likely to cause permanent injury to a person who handled it or who was otherwise in contact with it for more than a few minutes. It would probably be fatal to be close to this amount of unshielded radioactive material for a period in the range of a few minutes to an hour.

One member of the public is undergoing medical assessment in Mexico City after presenting himself with skin damage indicating overexposure to the source by carrying it over one shoulder, CNSNS said. A further 60–70 people have presented themselves for testing but have not shown signs of overexposure.

A person exposed to the source does not represent a contamination risk to others. Based on the information available, the Mexican authorities and the IAEA believe the general public is safe and will remain safe.

The IAEA remains in close contact with the Mexican authorities. It made an offer of good offices to Mexico when the source first went missing and remains ready to provide assistance if requested.

REFERENCES

[1] INTERNATIONAL ATOMIC ENERGY AGENCY, Convention on Early Notification of a Nuclear Accident, INFCIRC/335, IAEA, Vienna (1986).

[2] INTERNATIONAL ATOMIC ENERGY AGENCY, "Convention on Assistance in the Case of a Nuclear Accident or Radiological Emergency, INFCIRC/336, IAEA, Vienna (1986).

[3] FOOD AND AGRICULTURE ORGANIZATION OF THE UNITED NATIONS, INTERNATIONAL ATOMIC ENERGY AGENCY, INTERNATIONAL CIVIL AVIATION ORGANIZATION, INTERNATIONAL LABOUR ORGANIZATION, INTERNATIONAL MARITIME ORGANIZATION, INTERPOL, OECD NUCLEAR ENERGY AGENCY, PAN AMERICAN HEALTH ORGANIZATION, PREPARATORY COMMISSION FOR THE COMPREHENSIVE NUCLEAR-TEST-BAN TREATY ORGANIZATION, UNITED NATIONS ENVIRONMENT PROGRAMME, UNITED NATIONS OFFICE FOR THE COORDINATION OF HUMANITARIAN AFFAIRS, WORLD HEALTH ORGANIZATION, WORLD METEOROLOGICAL ORGANIZATION, Preparedness and Response for a Nuclear or Radiological Emergency, IAEA Safety Standards Series No. GSR Part 7, IAEA, Vienna (2015).

[4] NTERNATIONAL ATOMIC ENERGY AGENCY, Regulations for the Safe Transport of Radioactive Material, Specific Safety Requirement No. SSR-6 (Rev. 1), IAEA, Vienna (2018).

[5] INTERNATIONAL ATOMIC ENERGY AGENCY, Preparedness and Response for a Nuclear or Radiological Emergency Involving the Transport of Radioactive Material, SSG-65, IAEA, Vienna (2020).

[6] INTERNATIONAL ATOMIC ENERGY AGENCY, Security of Radioactive Material in Transport, Nuclear Security Series No. 9-G (Rev. 1), IAEA, Vienna (2020).

[7] EUROPEAN COMMISSION, FOOD AND AGRICULTURE ORGANIZATION OF THE UNITED NATIONS, INTERNATIONAL ATOMIC ENERGY AGENCY, INTERNATIONAL LABOUR ORGANIZATION, OECD NUCLEAR ENERGY AGENCY, PAN AMERICAN HEALTH ORGANIZATION, UNITED NATIONS ENVIRONMENT PROGRAMME, WORLD HEALTH ORGANIZATION, Radiation Protection and Safety of Radiation Sources: International Basic Safety Standards, IAEA Safety Standards Series No. GSR Part 3, IAEA, Vienna (2014).

[8] INTERNATIONAL ATOMIC ENERGY AGENCY, Code of Conduct on the Safety and Security of Radioactive Sources, IAEA, Vienna (2004).

[9] INTERNATIONAL ATOMIC ENERGY AGENCY, Categorization of Radioactive Sources, Safety Guide No. RS-G-1.9, IAEA, Vienna (2005).

[10] INTERNATIONAL ATOMIC ENERGY AGENCY, Dangerous quantities of radioactive materila (D-values), EPR-D-Values 2006, IAEA, Vienna (2006).

[11] FOOD AND AGRICULTURE ORGANIZATION OF THE UNITED NATIONS, INTERNATIONAL ATOMIC ENERGY AGENCY, INTERNATIONAL LABOUR OFFICE, PAN AMERICAN HEALTH ORGANIZATION, WORLD HEALTH ORGANIZATION, Criteria for Use in Preparedness and Response for a Nuclear or Radiological Emergency, IAEA Safety Standards Series No. GSG-2, IAEA, Vienna (2011).

[12] FOOD AND AGRICULTURE ORGANIZATION OF THE UNITED NATIONS, INTERNATIONAL ATOMIC ENERGY AGENCY, INTERNATIONAL LABOUR OFFICE, PAN AMERICAN HEALTH ORGANIZATION, UNITED NATIONS OFFICE FOR THE COORDINATION OF HUMANITARIAN AFFAIRS, WORLD HEALTH ORGANIZATION, Arrangements for Preparedness for a Nuclear or Radiological Emergency, IAEA Safety Standards Series No. GS-G-2.1, IAEA, Vienna (2007).

[13] FOOD AND AGRICULTURE ORGANIZATION OF THE UNITED NATIONS, INTERNATIONAL ATOMIC ENERGY AGENCY, INTERNATIONAL CIVIL AVIATION ORGANIZATION, INTERNATIONAL LABOUR OFFICE, INTERNATIONAL MARITIME ORGANIZATION, INTERPOL, OECD NUCLEAR ENERGY AGENCY, UNITED NATIONS OFFICE FOR THE COORDINATION OF HUMANITARIAN AFFAIRS, WORLD HEALTH ORGANIZATION, WORLD METEOROLOGICAL ORGANIZATION, Arrangements for the Termination of a Nuclear or Radiological Emergency, IAEA Safety Standards Series No. GSG-11, IAEA, Vienna (2018).

[14] FOOD AND AGRICULTURE ORGANIZATION OF THE UNITED NATIONS, INTERNATIONAL ATOMIC ENERGY AGENCY, INTERNATIONAL CIVIL AVIATION ORGANIZATION, INTERPOL, PREPARATORY COMMISSION FOR THE COMPREHENSIVE NUCLEAR-TEST-BAN TREATY ORGANIZATION AND UNITED NATIONS OFFICE FOR OUTER SPACE AFFAIRS, Arrangements for Public Communication in Preparedness and Response for a Nuclear or Radiological Emergency, IAEA Safety Standards Series No. GSG-14, IAEA, Vienna (2020).

[15] INTERNATIONAL ATOMIC ENERGY AGENCY, The Radiological Accident in Ventanilla, IAEA, Vienna (2019).

[16] INTERNATIONAL ATOMIC ENERGY AGENCY, The Radiological Accident in Chilca, IAEA, Vienna (2018).

[17] INTERNATIONAL ATOMIC ENERGY AGENCY, The Fukushima Daiichi Accident, IAEA, Vienna (2015).

[18] INTERNATIONAL ATOMIC ENERGY AGENCY, The Radiological Accident in Lia, Georgia, IAEA, Vienna (2014).

[19] INTERNATIONAL ATOMIC ENERGY AGENCY, The Radiological Accident in Nueva Aldea, IAEA, Vienna (2009).

[20] INTERNATIONAL ATOMIC ENERGY AGENCY, The Radiological Accident in Cochabamba, IAEA, Vienna (2004).

[21] INTERNATIONAL ATOMIC ENERGY AGENCY, Accidenttal Overexposure of Radiotherapy Patients in Bialystok, IAEA, Vienna (2004).

[22] INTERNATIONAL ATOMIC ENERGY AGENCY, The Radiological Accident in Gilan," IAEA, Vienna (2002).

[23] INTERNATIONAL ATOMIC ENERGY AGENCY, "The Radiological Accident in Samut Prakarn," IAEA, Vienna (2002).

[24] INTERNATIONAL ATOMIC ENERGY AGENCY, The Criticality Accident in Sarov, IAEA, Vienna (2001).

[25] INTERNATIONAL ATOMIC ENERGY AGENCY, The Radiological Accident in Lilo, IAEA, Vienna (2000).

[26] INTERNATIONAL ATOMIC ENERGY AGENCY, The Radiological Accident in Istanbul, IAEA, Vienna (2000).

[27] INTERNATIONAL ATOMIC ENERGY AGENCY, The Radiological Accident in Yanango, IAEA, Vienna (2000).

[28] INTERNATIONAL ATOMIC ENERGY AGENCY, The Radiological Accident in the Processing Plant at Tomsk, IAEA, Vienna (1998).

[29] INTERNATIONAL ATOMIC ENERGY AGENCY, The Radiological Accident in Tammiku, IAEA, Vienna (1998).

[30] INTERNATIONAL ATOMIC ENERGY AGENCY, Accidental Overexposure of Radiotherapy Patients in San Jose, Costa Rica, IAEA, Vienna (1998).

[31] INTERNATIONAL ATOMIC ENERGY AGENCY, The Radiological Accident at the Irradiaiton Facility in Nesvizh, IAEA, Vienna (1996).

[32] INTERNATIONAL ATOMIC ENERGY AGENCY, An Electron Acceleratro Accident in Hanoi, Vietnam, IAEA, Vienna (1996).

[33] INTERNATIONAL ATOMIC ENERGY AGENCY, The Radiological Accident in Soreq, IAEA, Vienna (1993).

[34] INTERNATIONAL ATOMIC ENERGY AGENCY, The Radiological Accident in San Salvador, IAEA, Vienna (1990).

[35] INTERNATIONAL ATOMIC ENERGY AGENCY, The Radiological Accident in Goiania, IAEA, Vienna (1988).

[36] INTERNATIONAL ATOMIC ENERGY AGENCY, IAEA Safety Glossary, Terminology Used in Nuclear Safety and Radiation Protection 2018 Edition, IAEA, Vienna (2018)

[37] NTERNATIONAL ATOMIC ENERGY AGENCY, Report of the IRRS (Integrated Regulatory Review Service) Mission to Mexico, 26 November to 5 December 2007, IAEA, Vienna (2007).

[38] https://www.dof.gob.mx/nota_detalle.php?codigo=5479227&fecha=10/04/2017#gsc.ta
b=0

[39] INTERNATIONAL ATOMIC ENERGY AGENCY, Operations Manual for Incident and Emergencies, EPR-IEComm, IAEA, Vienna (2019).

[40] INTERNATIONAL ATOMIC ENERGY AGENCY, IAEA Response and Assistance Network, EPR-RANET, IAEA, Vienna (2018).

[41] INTERNATIONAL ATOMIC ENERGY AGENCY, INES (The International Nuclear and Radiological Event Scale) User's Manual, IAEA, Vienna (2008).

CONTRIBUTORS TO DRAFTING AND REVIEW

Alcantara, O.	Comisión Nacional de Seguridad Nuclear y Salvaguardias
Baciu, F.	International Atomic Energy Agency
Buglova, E.	International Atomic Energy Agency
Cabrera, O.	Comisión Nacional de Seguridad Nuclear y Salvaguardias
Caria, C.	International Atomic Energy Agency
Jayarajan, N.	International Atomic Energy Agency
Hussain, M.	International Atomic Energy Agency
Kaiser, P.	International Atomic Energy Agency
Kasper, M.	International Atomic Energy Agency
Ladsous, D.	International Atomic Energy Agency
Nestoroska Madjunarova, S.	International Atomic Energy Agency
Nikolaki, M.	International Atomic Energy Agency
Pacheco Jimenez, R.	International Atomic Energy Agency
Reyes, H.	Comisión Nacional de Seguridad Nuclear y Salvaguardias
Romero, C.	Comisión Nacional de Seguridad Nuclear y Salvaguardias
Smith, K.	International Atomic Energy Agency
Torres Vidal, C.	International Atomic Energy Agency
Vargas, K.	International Atomic Energy Agency

IAEA
International Atomic Energy Agency

ORDERING LOCALLY

IAEA priced publications may be purchased from the sources listed below or from major local booksellers.

Orders for unpriced publications should be made directly to the IAEA. The contact details are given at the end of this list.

NORTH AMERICA

Bernan / Rowman & Littlefield

15250 NBN Way, Blue Ridge Summit, PA 17214, USA

Telephone: +1 800 462 6420 • Fax: +1 800 338 4550

Email: orders@rowman.com • Web site: www.rowman.com/bernan

REST OF WORLD

Please contact your preferred local supplier, or our lead distributor:

Eurospan Group

Gray's Inn House
127 Clerkenwell Road
London EC1R 5DB
United Kingdom

Trade orders and enquiries:

Telephone: +44 (0)176 760 4972 • Fax: +44 (0)176 760 1640
Email: eurospan@turpin-distribution.com

Individual orders:

www.eurospanbookstore.com/iaea

For further information:

Telephone: +44 (0)207 240 0856 • Fax: +44 (0)207 379 0609
Email: info@eurospangroup.com • Web site: www.eurospangroup.com

Orders for both priced and unpriced publications may be addressed directly to:

Marketing and Sales Unit
International Atomic Energy Agency
Vienna International Centre, PO Box 100, 1400 Vienna, Austria
Telephone: +43 1 2600 22529 or 22530 • Fax: +43 1 26007 22529
Email: sales.publications@iaea.org • Web site: www.iaea.org/publications

22-03497E